,,,

*50 Great Flying and Underwater Perpetual Motion
Machines*

*Dedicated to the idealists, without whom
perpetual motion would be impossible.*

50 Great Flying and Underwater

Perpetual Motion Machines

by Nathan Coppedge

50 Great Flying and Underwater Perpetual Motion Machines

,,,

50 Great Flying and Underwater Perpetual Motion Machines

Preface / Intro

This is the companion volume to 100 Great Perpetual Motion Machines. In this text are found many visual descriptions of certain types of water devices, amphibious elements, and flying apparatuses, which have a reputation for possibly working. Many of the inventions are of the author's own creation. Also included are a number of inventions attributed to friends of the family.

50 Great Flying and Underwater Perpetual Motion
Machines

,,,

Amphibious Vehicle, Underwater Buoy Device 2

UNDERWATER BUOY DEVICE 2

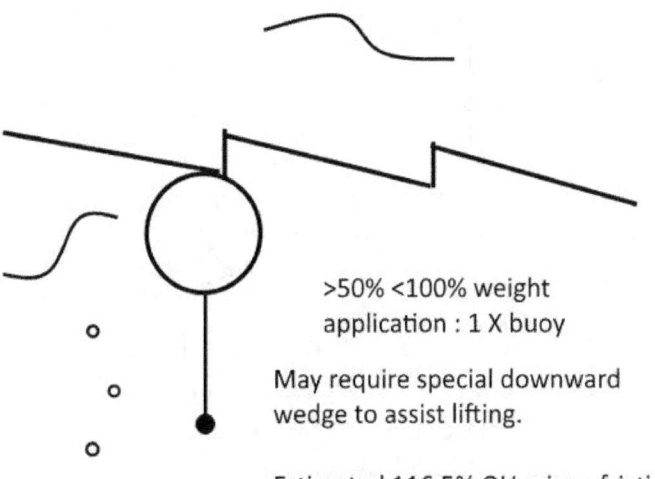

>50% <100% weight application : 1 X buoy

May require special downward wedge to assist lifting.

Estimated 116.5% OU minus friction

Also, there may be a helium variation of the Underwater Escher Machine

Anti-Machine Variation 1

NON-EXPONENTIAL EFFICIENCY
"ANTI-KNOWLEDGE" ITERATION 1

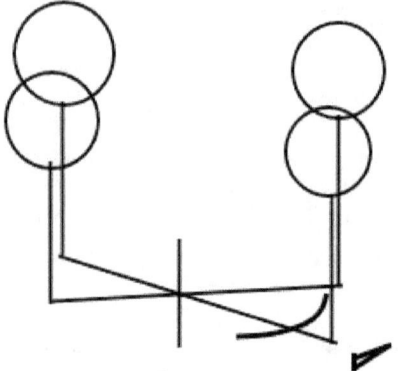

TRACK ELEMENT IS
ABOVE FULCRUM LEVEL
BALLOON OPPOSITE TRACK ELEMENT
CAUSES DOWNWARD ROTATION
OF OPPOSITE END DUE TO
UPPER ABRASION CLOSE TO
EQUILIBRIUM ON OPPOSITE END.
THE LOWERING OF ONE END
IS THEORETICALLY RECOVERABLE
DUE TO EQUILIBRIZATION.
PERPENDICULAR BALLOONS ARE
IN ROUGH EQUILIBRIUM, BUT
STILL ROTATE.

NATHAN LARKIN COPPEDGE
MARCH 13, 2023

Anti-Machine Variation 2

"ANTI-MACHINE" ITERATION 1

NATHAN LARKIN COPPEDGE
MARCH 13, 2023

1X BALLOON

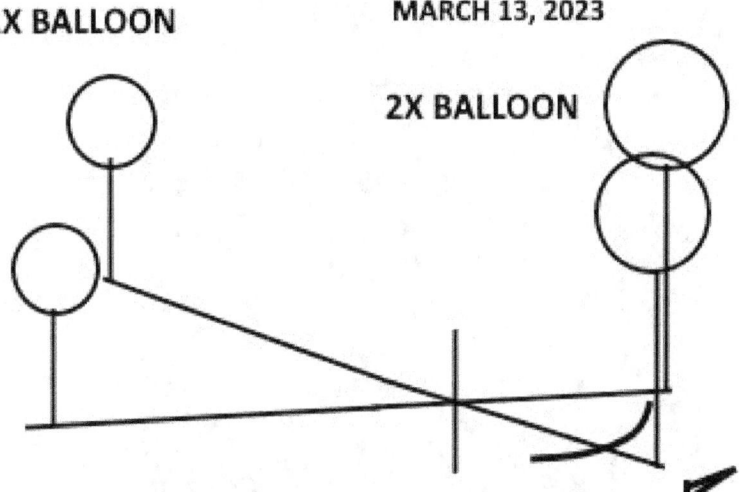

2X BALLOON

BEAMS ARE WEIGHTED TO BALANCE
APART FROM BALLOONS.
TRACK ELEMENT IS DOWNWARDS SLOPED,
AND IS POSITIONED INITIALLY ABOVE
THE LEVER ON THE SHORT END.
1X BALLOON USES LEVERAGE TO PULL
LARGER BALLOON DOWNWARDS, USING
SUPPORT FROM BAR TO REDUCE
BUOYANCY OF 2X BALLOON.

Auto Elevator

Automatic Rowing

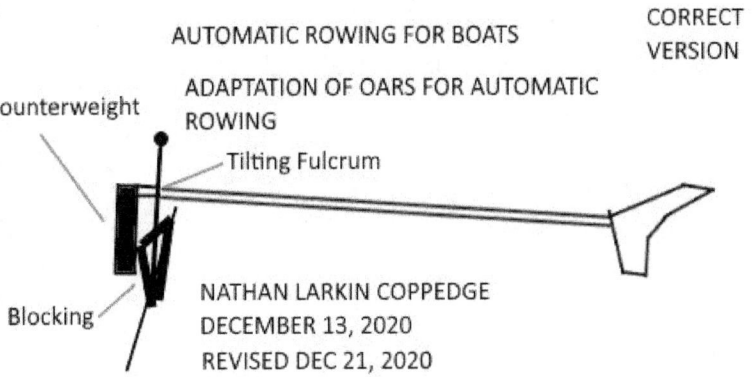

AUTOMATIC ROWING FOR BOATS

CORRECT VERSION

ADAPTATION OF OARS FOR AUTOMATIC ROWING

counterweight

Tilting Fulcrum

Blocking

NATHAN LARKIN COPPEDGE
DECEMBER 13, 2020
REVISED DEC 21, 2020

Tilting the fulcrum forward creates forward movement.
Tilting the fulcrum backward returns the oar.

Blocking may be changed to allow return at higher altitude.

Effective Rowing: Improvement:

- Flanges would be fitted to the poles.
- Rowing would take place by shifting the lever backwards all the way, causing the pole to shoot upwards.
- Additional smaller back-positioned counterweights would lift the fulcrum upwards slightly as the lever is shifted backwards.
- The forward motion of the lever would drop the flanges simultaneously forwards with limited resistance due to

the medium counterweight and allow the oar to drag a long distance backwards, due to the placement of the net and motion traveled during the drop.

Boat, 1ˢᵗ Generation

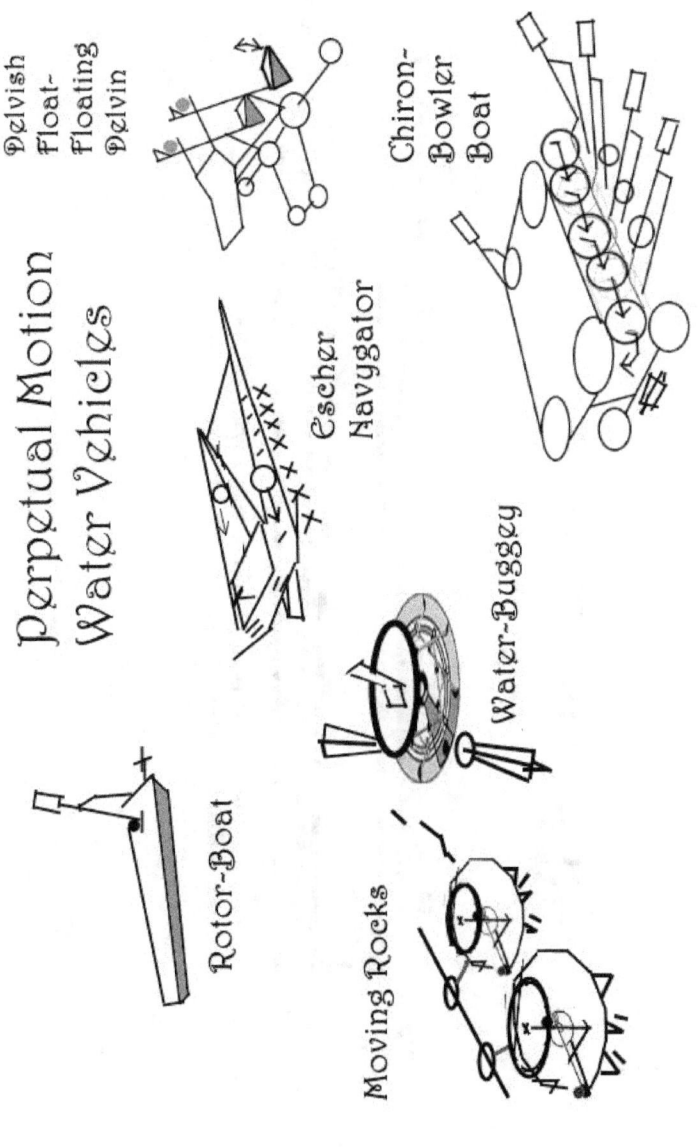

Perpetual Motion Water Vehicles

Delvish Float-Floating Delvin

Chiron-Bowler Boat

Escher Navygator

Rotor-Boat

Water-Buggey

Moving Rocks

Nathan Coppedge 2019/06/12

Boat, 2nd Generation, Free Energy

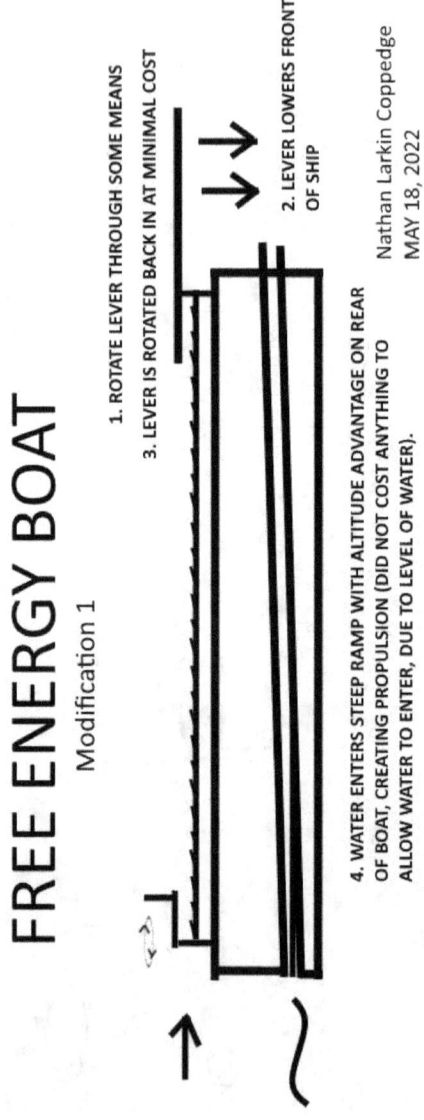

FREE ENERGY BOAT

Modification 1

1. ROTATE LEVER THROUGH SOME MEANS

3. LEVER IS ROTATED BACK IN AT MINIMAL COST

2. LEVER LOWERS FRONT OF SHIP

4. WATER ENTERS STEEP RAMP WITH ALTITUDE ADVANTAGE ON REAR OF BOAT, CREATING PROPULSION (DID NOT COST ANYTHING TO ALLOW WATER TO ENTER, DUE TO LEVEL OF WATER).

Nathan Larkin Coppedge
MAY 18, 2022

Boats, 3rd Generation

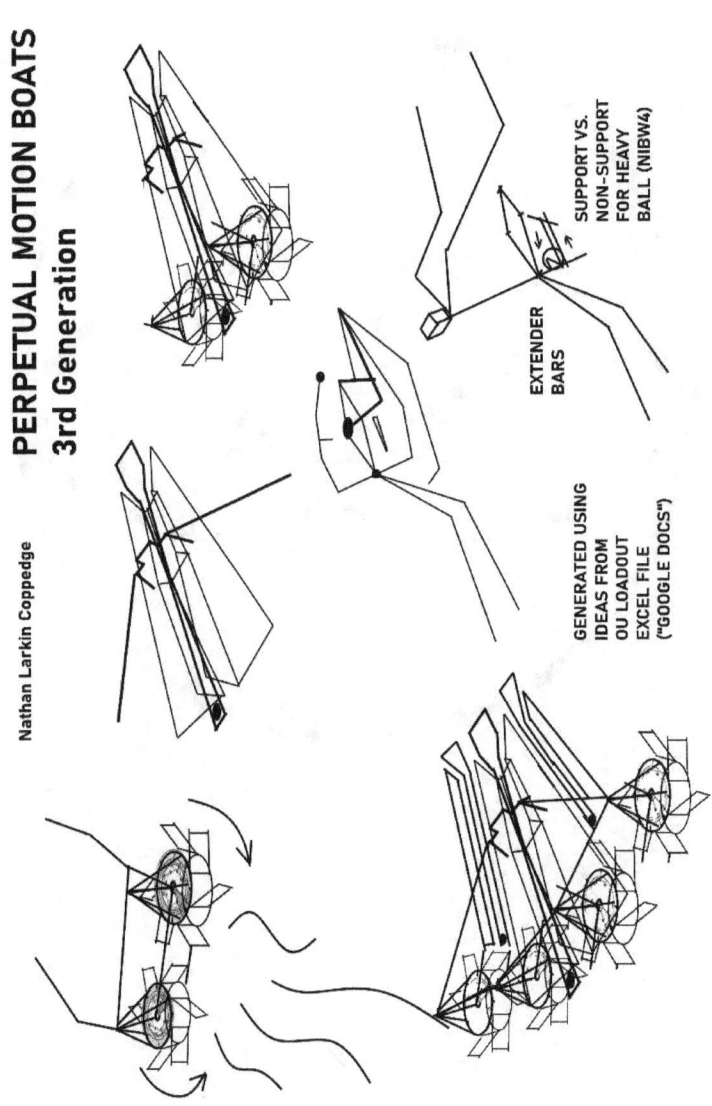

PERPETUAL MOTION BOATS
3rd Generation

Nathan Larkin Coppedge

SUPPORT VS. NON-SUPPORT FOR HEAVY BALL (NIBW4)

EXTENDER BARS

GENERATED USING IDEAS FROM OU LOADOUT EXCEL FILE ("GOOGLE DOCS")

Bobblety Toy, Helium

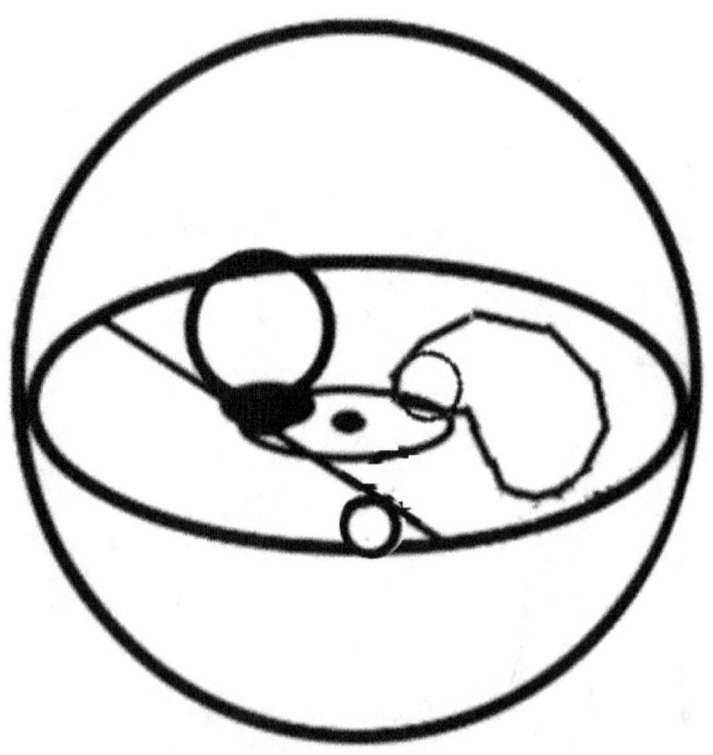

Bobbleties, Flying: Additional

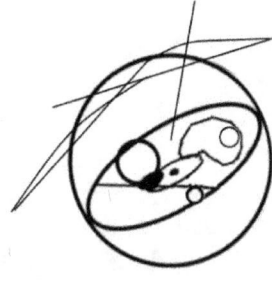 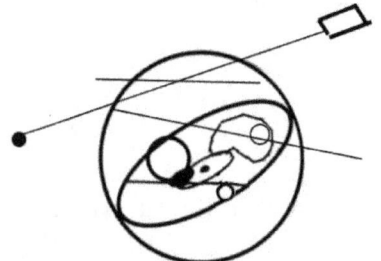

HELIUM ESCHER
BOBBLETY

HELIUM WEDGE & CTRWGHT
BOBBLETY

Buoyant Lever

Submarines and balloons are known to be able to use a long pole to cause descent which does not occur when the pole is balanced across the center axis.

Buoy-Lever

A version of the buoyancy lever, the basis for a likely flying machine principle (an earlier experiment was run on July 16, 2020):

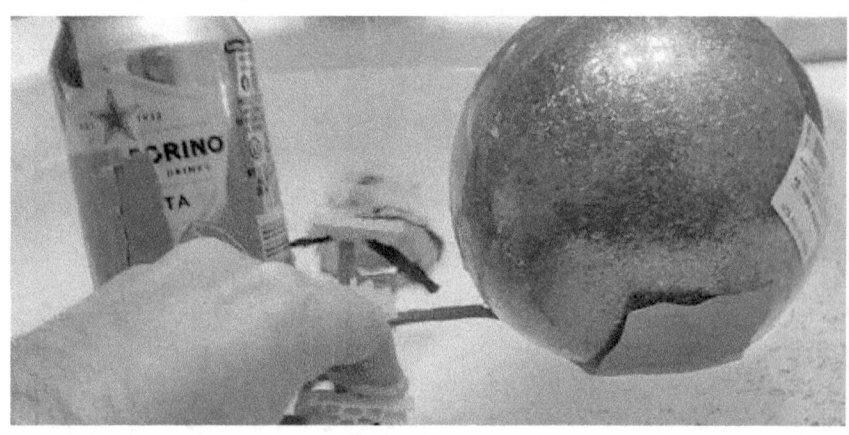

...

Buoy-Wheel with Unbalanced Chains

BUOY WHEEL USING
UNBALANCED CHAINS

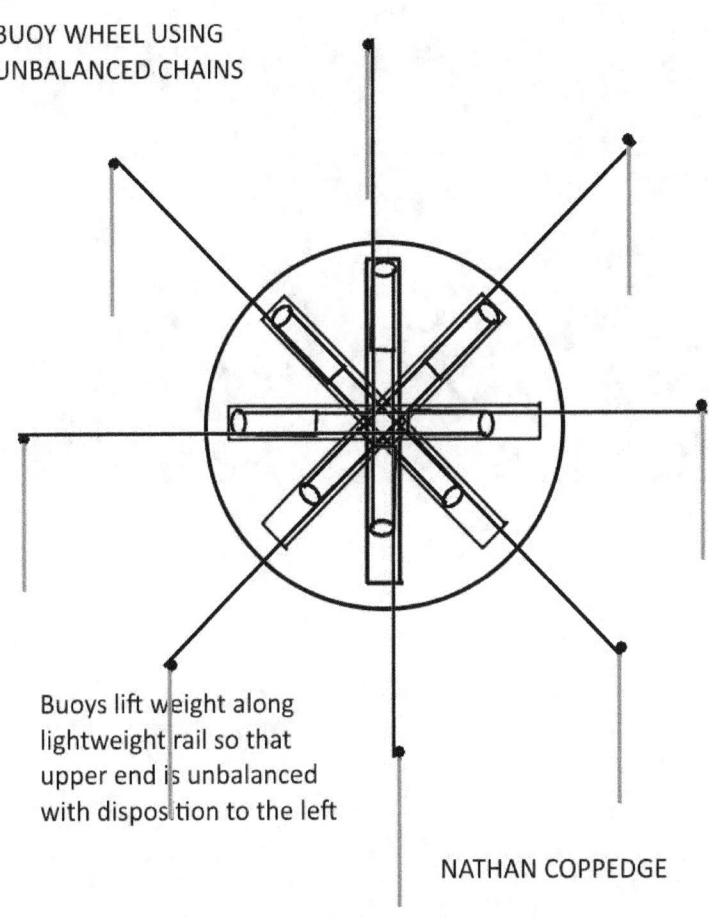

Buoys lift weight along
lightweight rail so that
upper end is unbalanced
with disposition to the left

NATHAN COPPEDGE

Capillary Action

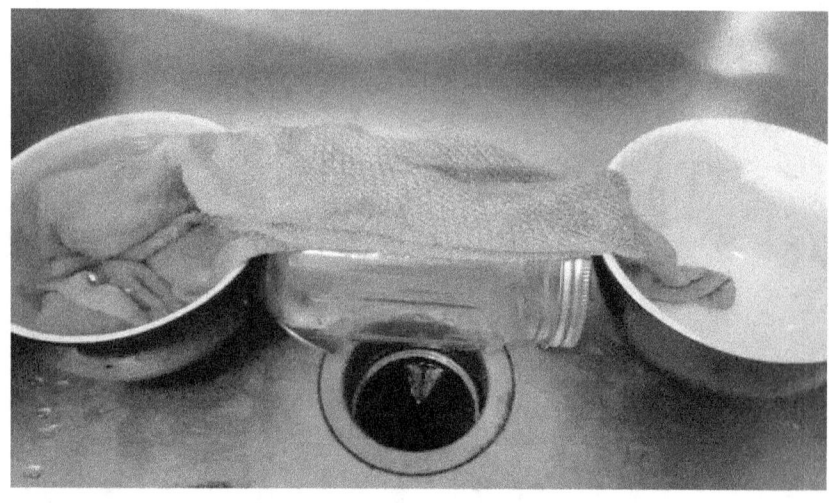

Counterbuoy Spiral Devices

BUOY SPIRALS AND SOME OTHER UNDERWATER DEVICES

CTRBUOY SPIR _ EXEFF2	CTRBUOY SPIR _ EXEFF1	CTRBUOY BUOY SPIR _	DIFF BUOY SPIRAL _	(other)
CTRBUOY SPIR 5 EXEFF2 — EFF = -5, DIFF = 5, RESULTS	**CTRBUOY SPIR 5 EXEFF1** — EFF = -4, DIFF = 5, RESULTS > -7 < -5	**CTRBUOY BUOY SPIR 5** — EFF = -3, DIFF = 5, RESULTS > -6 < -4.5	**DIFF BUOY SPIRAL 5** — EFF = -2, DIFF = 5, RESULTS > -5 < -4	**REPEAT BUOY 4.2** — EFF = -1, DIFF = 5, RESULTS > -4 < -3.5
CTRBUOY SPIR 4 EXEFF2 — EFF = -5, DIFF = 4, RESULTS	**CTRBUOY SPIR 4 EXEFF1** — EFF = -4, DIFF = 4, RESULTS > -6 < -4	**CTRBUOY BUOY SPIR 4** — EFF = -3, DIFF = 4, RESULTS > -5 < -3.5	**DIFF BUOY SPIRAL 4** — EFF = -2, DIFF = 4, RESULTS > -4 < -3	**IMPROVED VERT BUOY** — EFF = -1, DIFF = 4, RESULTS > -3 < -2.5
CTRBUOY SPIR 3 EXEFF2 — EFF = -5, DIFF = 3, RESULTS	**CTRBUOY SPIR 3 EXEFF1** — EFF = -4, DIFF = 3, RESULTS > -5 < -3	**CTRBUOY BUOY SPIR 3** — EFF = -3, DIFF = 3, RESULTS > -4 < -2.5	**DIFF BUOY SPIRAL 3** — EFF = -2, DIFF = 3, RESULTS > -3 < -2	**VERTICAL BUOY** — EFF = -1, DIFF = 3, RESULTS > -2 < -1.5
CTRBUOY SPIR 2 EXEFF2 — EFF = -5, DIFF = 2, RESULTS	**CTRBUOY SPIR 2 EXEFF1** — EFF = -4, DIFF = 2, RESULTS > -4 < -2	**CTRBUOY BUOY SPIR 2** — EFF = -3, DIFF = 2, RESULTS > -3 < -1.5	**DIFF BUOY SPIRAL 2** — EFF = -2, DIFF = 2, RESULTS > -2 < -1	**SWIVEL BUOY** — EFF = -1, DIFF = 2, RESULTS > -1 < -0.5
CTRBUOY SPIR 1 EXEFF 2 — EFF = -5, DIFF = 1, RESULTS	**CTRBUOY SPIR 1 EXEFF 1** — EFF = -4, DIFF = 1, RESULTS > -3 < -1	**CTRBUOY BUOY SPIR 1** — EFF = -3, DIFF = 1, RESULTS > -2 < -0.5	**DIFF BUOY SPIRAL 1** — EFF = -2, DIFF = 1, RESULTS > -1 < 0	**BALLOON 2 PMM** — EFF = -1, DIFF = 1, RESULTS > 0 < 0.5

Crescent Weighted Balloon Device

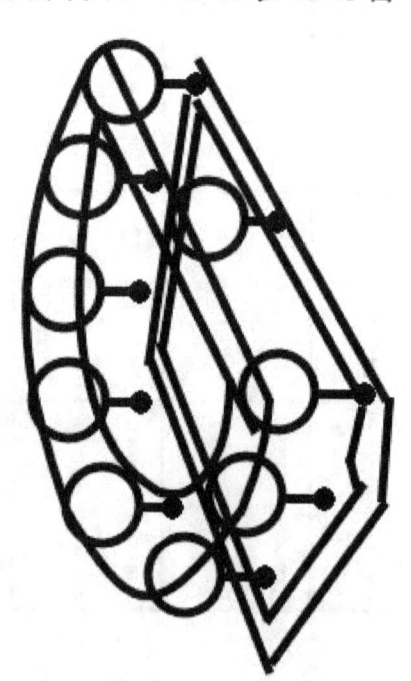

BUOYANCY OF EACH
BALLOON = 1X
DANGLING WEIGHT =
> 0.5 < 1X
(USING UPSIDE-DOWN
0.5 MASS X DISTANCE
FOR DESCENT)
WEIGHT IS SUPPORTED
SLIGHTLY USING
STANDARD 0.5 M X D
DURING UPWARD MOTION

Nathan Coppedge 2023-07-29

CRESCENT WEIGHTED BALLOON DEVICE
USING 0.5 MASS X DISTANCE

Flying Machines, 1st Generation: Arkites (?)

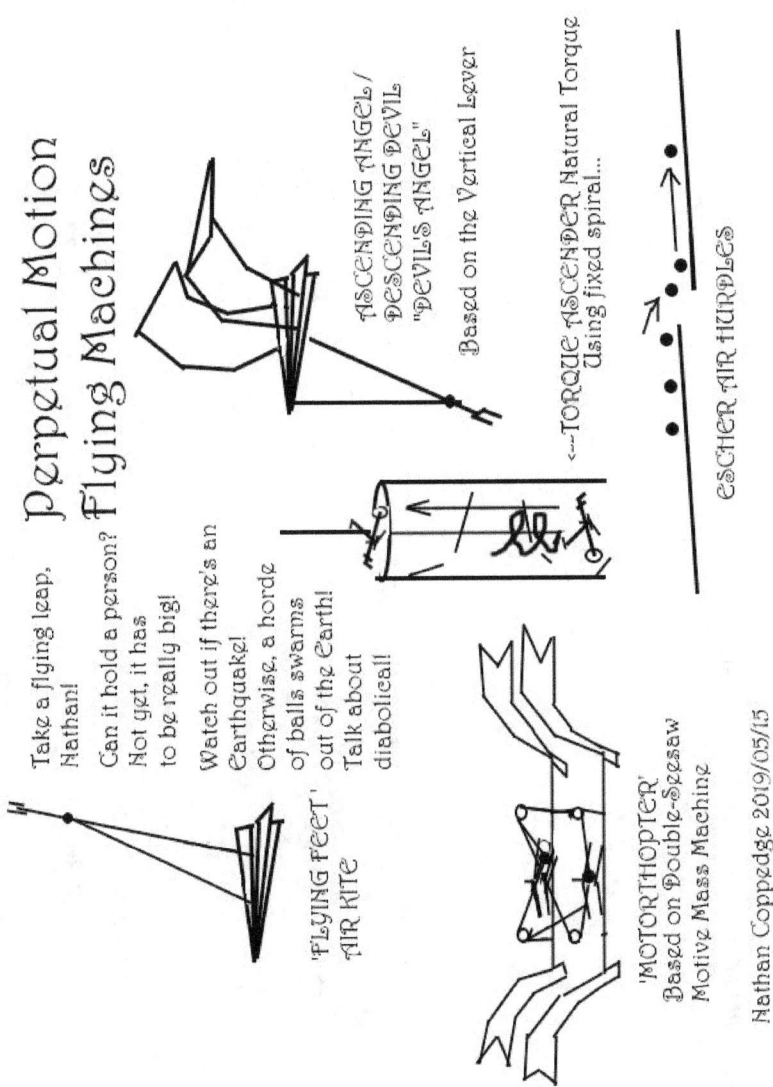

Flying Machines, 2nd Generation: Wanderers (?)

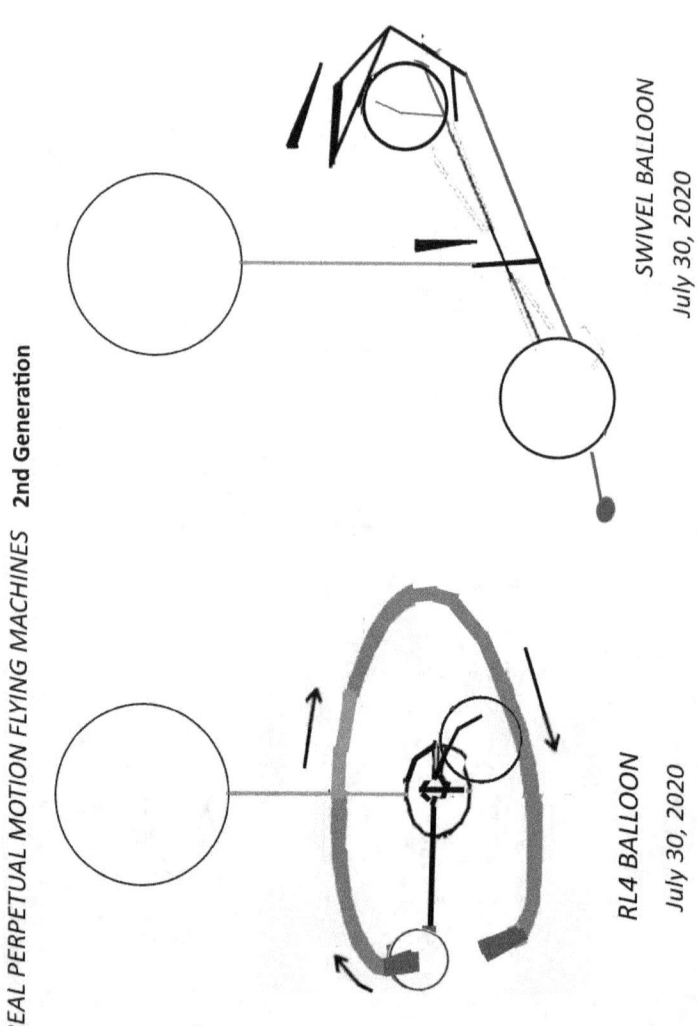

REAL PERPETUAL MOTION FLYING MACHINES   2nd Generation

SWIVEL BALLOON
July 30, 2020

RL4 BALLOON
July 30, 2020

NATHAN COPPEDGE

Flying Machines, 3ʳᵈ Generation

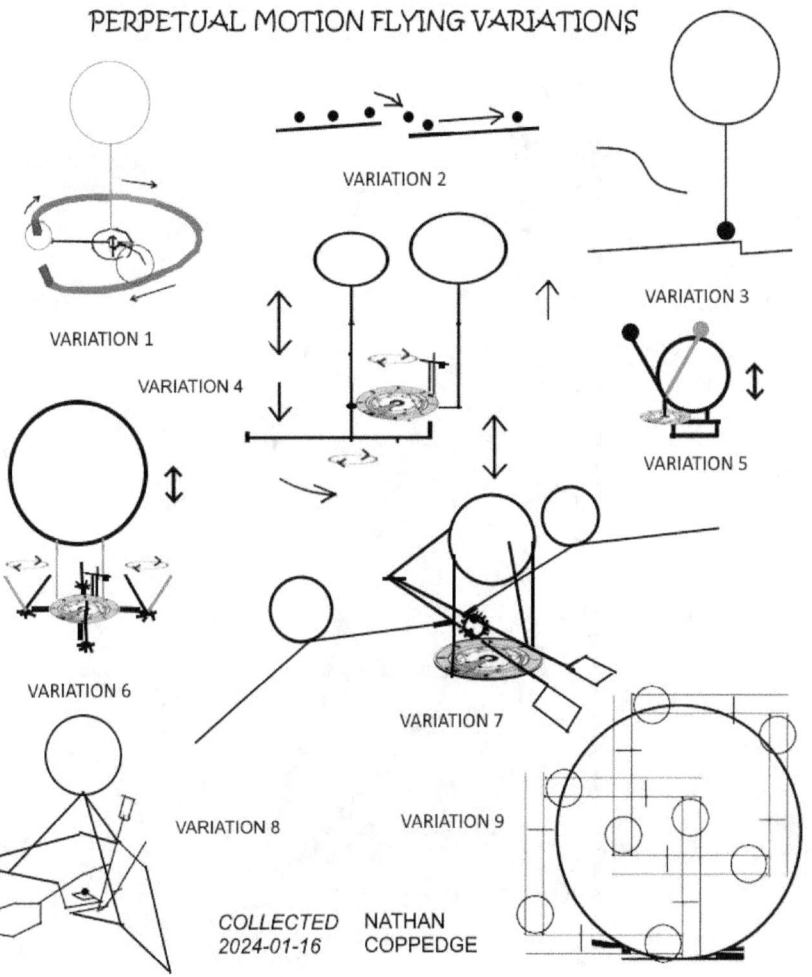

PERPETUAL MOTION FLYING VARIATIONS

VARIATION 1

VARIATION 2

VARIATION 3

VARIATION 4

VARIATION 5

VARIATION 6

VARIATION 7

VARIATION 8

VARIATION 9

COLLECTED   NATHAN
2024-01-16   COPPEDGE

It is thought the tilt motors could be used with a separate connection to power a tail giving complete mobility.

Frank Tatay's Buoy Device, Improved

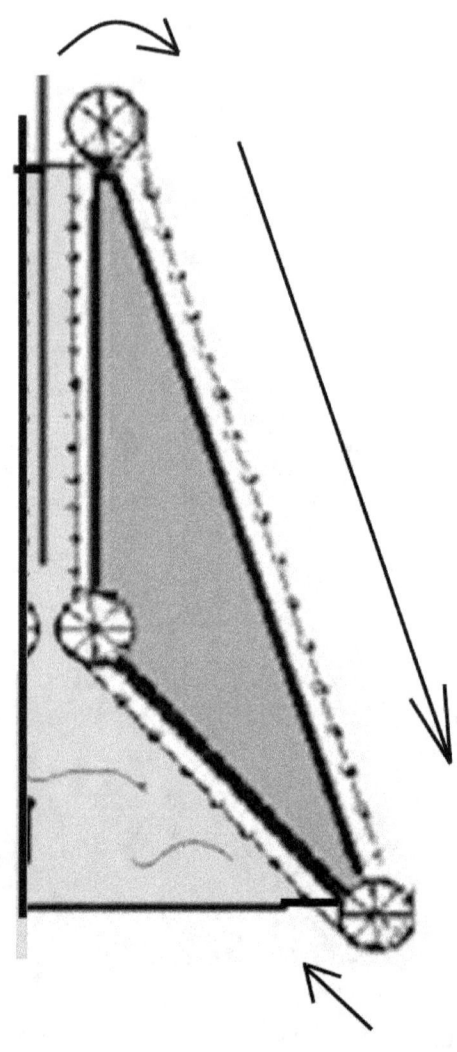

50 Great Flying and Underwater Perpetual Motion Machines

The latest stats seem to say this particular device doesn't work, however, there may be horizontal variations that have weak entry resistance or which take better advantage of falling weights perhaps using wheels with unbalanced leverage.

Free Elevator Concept

Due to the size of each of the balloons, turning the lever attached to the elevator towards the larger balloon causes lift, and towards the smaller balloon causes drag.

FREE ELEVATOR CONCEPT

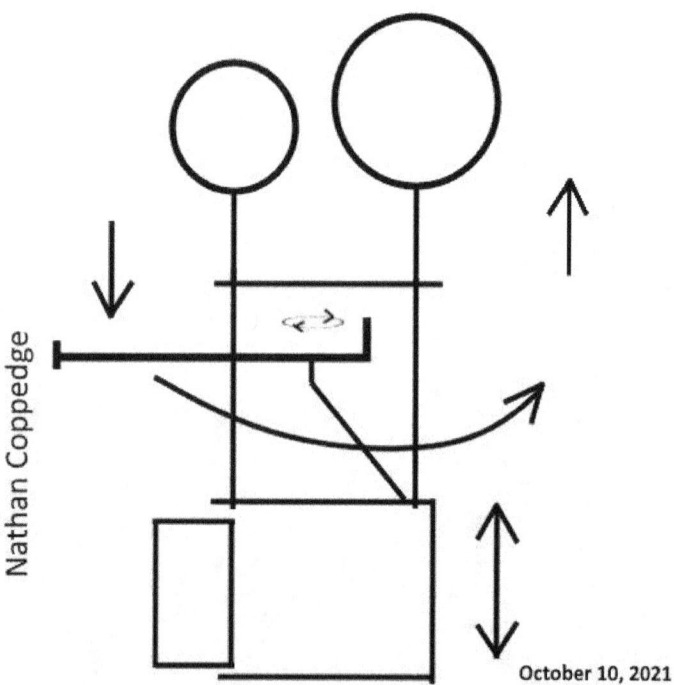

ABOVE: Free Elevator (Oct 10, 2021).

50 Great Flying and Underwater Perpetual Motion Machines

Sealed habitations could rise to the surface at no cost and sink down at no cost. Careful to consult divers about this stuff, there are serious painful conditions that can develop when rising quickly in water. In air, it is less dangerous, as I realized.

50 Great Flying and Underwater Perpetual Motion Machines

Genie-Buoy Device

First device I have found with the theoretical maximum rating of 200% for a ground-based device not using enhanced effects:

THE GENIE-BUOY DEVICE

Approx 18 buoys
(18/2) / (18/2) + 1 * 100
= < 200% OU

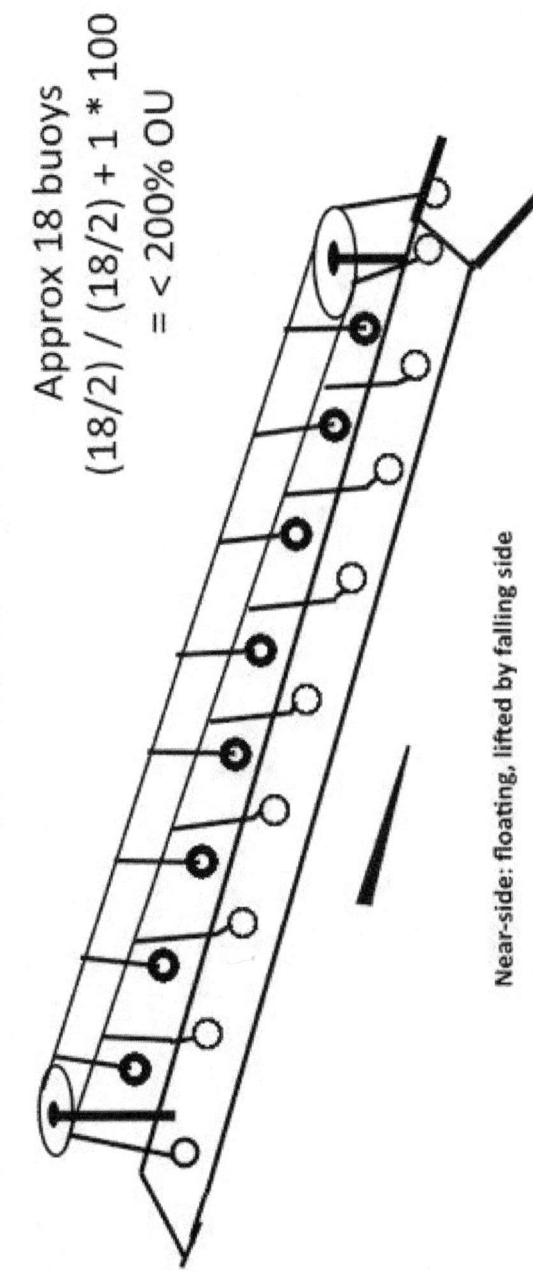

Near-side: floating, lifted by falling side

ILLUSTRATION BY NATHAN LARKIN COPPEDGE. Original concept may be due in this case to Jorge Vargas.

Buoys: Approx 18

Ratio: About 1 : 1 Equivalency.

OU Rating: < 200% Conventional OU

Principle: Reduced resistance with minimal entry resistance. Modular by the earliest functional definition: massive overbalanced process when in rare cases maybe it can be unbalanced. For example, brilliant solution this time, when the device is horizontally disposed, with half of the masses buoyant and not submerged.

Helium Balloon Perpetual Motion 1

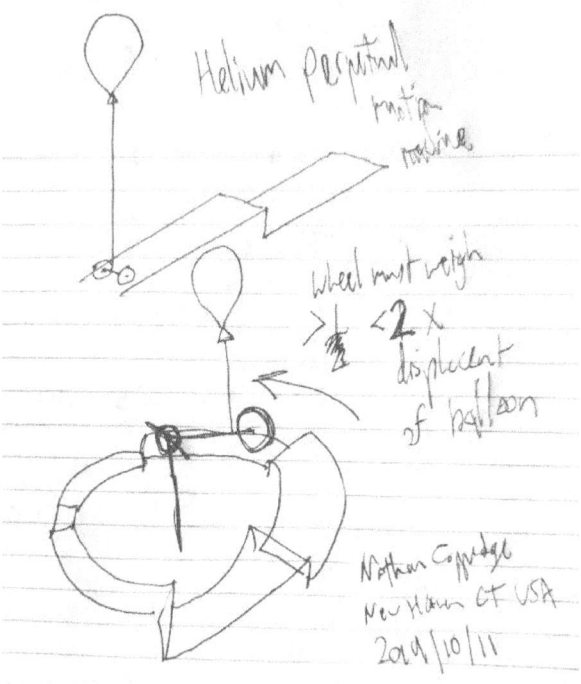

- Date of Invention: Presumably October 11, 2019.
- Notable Difference: Repeating circular Sawblade-shaped track allows wheel (s) or roller to follow path of least resistance. This design is now thought only to work when also combined with Escher Principle.
- State of affairs: Evidence is tentative, except that the relatively successful yet

long-running experiment on the Helium Balloon 2 suggests this also works. The problem with Helium 2 might be explained by the friction involved in the experiment.

- Over-Unity Rating: < 116.5% conventional Over-Unity.
- Buoyant Force: 1X
- Additional Weight: >1 to < 2 (>1 to < 1.5 recommended, with optimum around 1.3 to 1.4)
- Maximum Gradient: < 7.425 degrees.

Helium Balloon Iteration 2

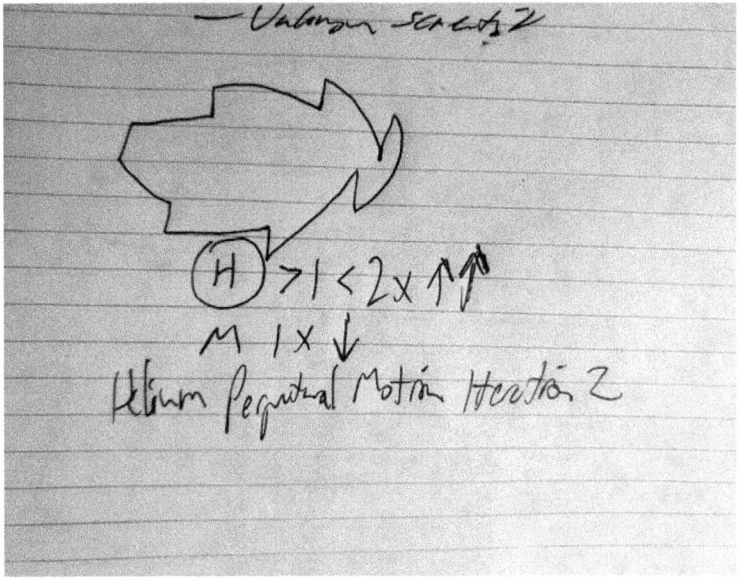

- Classification: HBall2
- Date of Invention: 1968 (guess based on numerology)
- Attribution: "I know a perpetual motion machine, but it requires a ridiculous amount of helium." --Larry Larkin, Coppedge's maternal grandfather
- State of affairs: This design has evidence if wind can be ruled out, although motion of the balloon was unbearably slow. However, since motion was consistent it is unlikely to have been variable wind.

- Rating: < 116.5% conventional Over-Unity (assuming 133% Buoyancy).
- Buoyant Force: > 1 < 2 X (suspected maximized around 1.33X).
- Additional Weight: 1X
- Maximum Gradient: < 7.425 degrees.

Helium Balloon 3

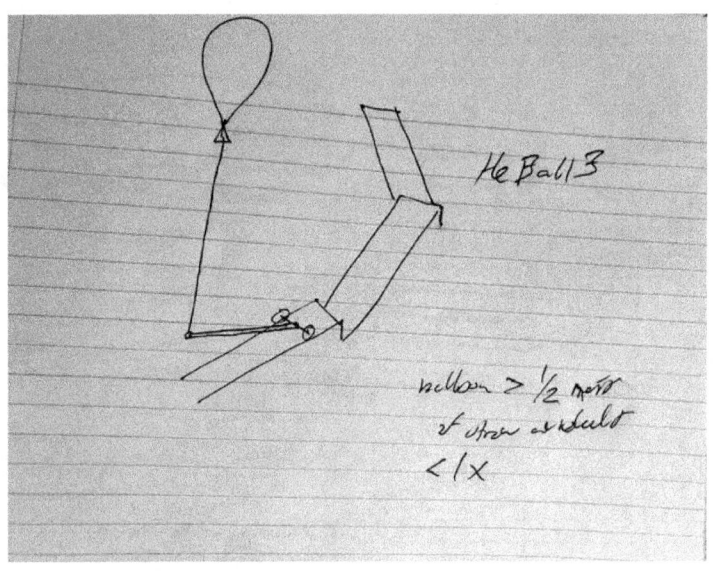

- Date of Invention: by November 2, 2019.
- Over-Unity Rating: < 116.5%
- Buoyant Force: 1X
- Additional Weight: >1 to < 2 (>1 to < 1.5 recommended, with optimum around 1.3 to 1.4)
- Maximum Gradient: < 7.425 degrees.

Helium Tilt Motor

Invented: July 2023

This is a version of the Tilt Motor that applies buoyancy upside down, with each lever operated on the short end by a buoy. Otherwise, it is seen to be identical to the Tilt Motor.

Jen's Bubblevator

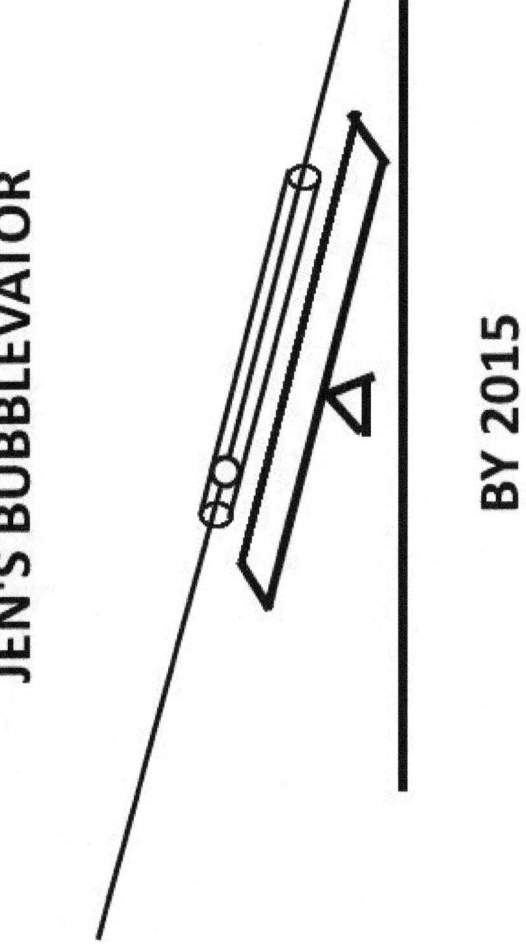

Rating: < 116.67 % Conventional OU

"Spaghetti" Leverage: 1 to 1.5 : 1

Buoyancy Range: >2.75X <3X

Additional "spaghetti" mass: 2X

Estimated Maximum gradient: < 7.425 degrees.

Jen's Buoyant Flange

SELF-ROTATING BUOYANT FLANGE

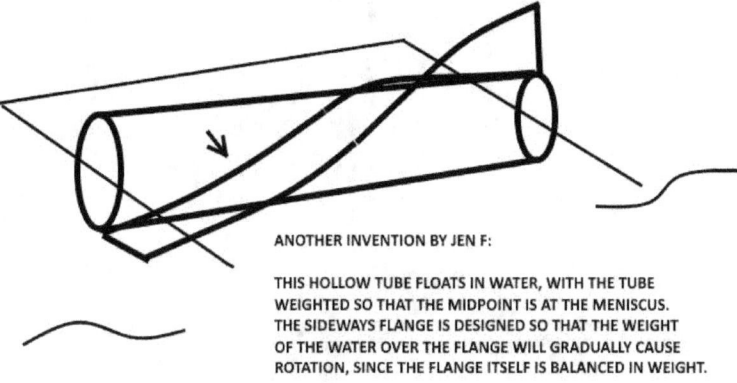

ANOTHER INVENTION BY JEN F:

THIS HOLLOW TUBE FLOATS IN WATER, WITH THE TUBE
WEIGHTED SO THAT THE MIDPOINT IS AT THE MENISCUS.
THE SIDEWAYS FLANGE IS DESIGNED SO THAT THE WEIGHT
OF THE WATER OVER THE FLANGE WILL GRADUALLY CAUSE
ROTATION, SINCE THE FLANGE ITSELF IS BALANCED IN WEIGHT.

Katy Ruben's Improved Water Wheel

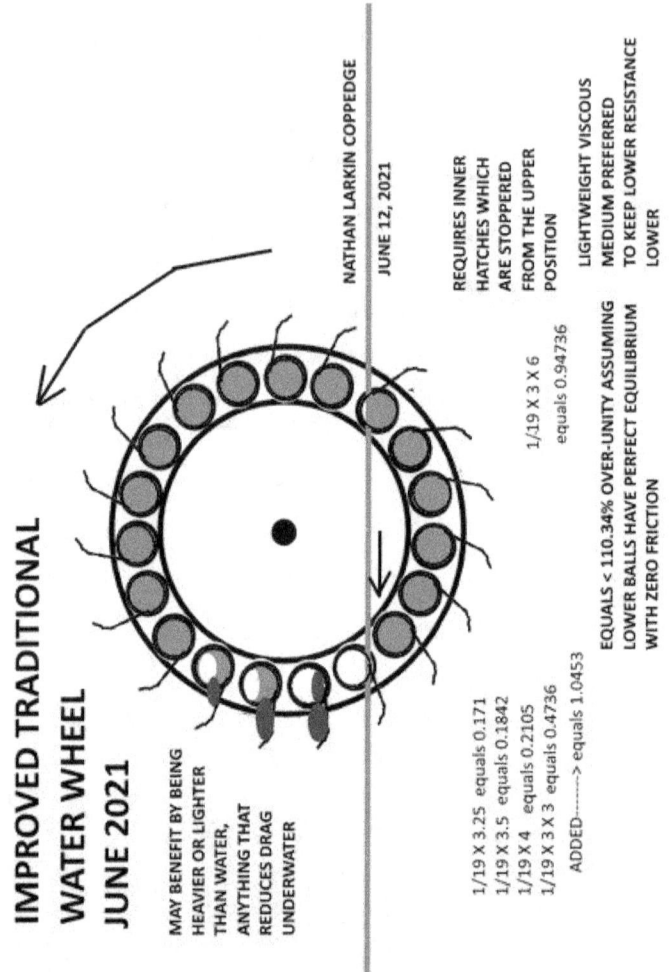

IMPROVED TRADITIONAL WATER WHEEL JUNE 2021

MAY BENEFIT BY BEING HEAVIER OR LIGHTER THAN WATER, ANYTHING THAT REDUCES DRAG UNDERWATER

NATHAN LARKIN COPPEDGE

JUNE 12, 2021

REQUIRES INNER HATCHES WHICH ARE STOPPERED FROM THE UPPER POSITION

LIGHTWEIGHT VISCOUS MEDIUM PREFERRED TO KEEP LOWER RESISTANCE LOWER

1/19 X 3 X 6
equals 0.94736

EQUALS < 110.34% OVER-UNITY ASSUMING LOWER BALLS HAVE PERFECT EQUILIBRIUM WITH ZERO FRICTION

1/19 X 3.25   equals 0.171
1/19 X 3.5   equals 0.1842
1/19 X 4   equals 0.2105
1/19 X 3 X 3 equals 0.4736
ADDED-------> equals 1.0453

- 0.94736 : 1.0453 resistance with balance
- Rating: 110.34% OU

Lever-Assisted Helium Balloons 1 and 2

This device is designed to use a counterbuoyed lever which is periodically lifted by a balloon which has a dangling weight. The amount of buoyancy varies depending on the design, however it is usually given as 0.5 to 1 to 1 favoring either the balloon or the dangling weight. The lever is supposed to provide additional efficiency by in this case assisting downward movement.

Mobila Planete, Mobile Spheres / Perpetual Motion Planets

"MOBILE SPHERE" PERPETUAL MOTION PLANET

Design #1

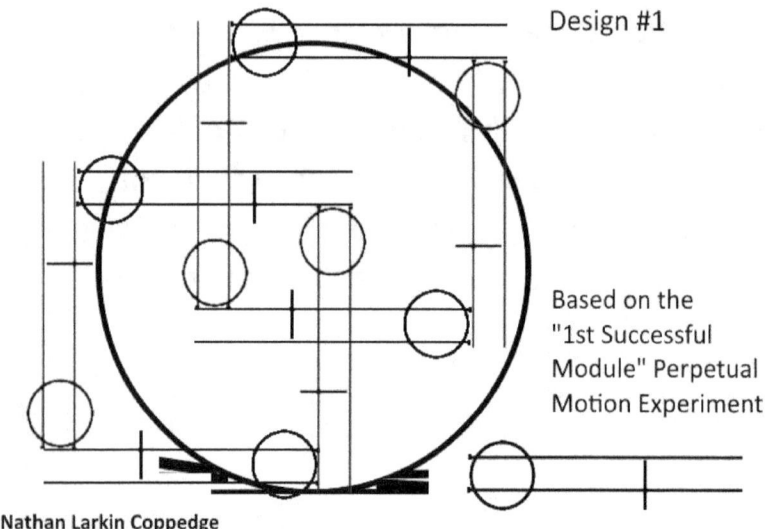

Based on the "1st Successful Module" Perpetual Motion Experiment

Nathan Larkin Coppedge

Date of Invention: Nov 30, 2020
Rating: <150% conventional Over-Unity minus mass.
Leverage in each machine: 1:1 (judging by counterweight distance)
Counterweight Mass: >1.5X to <2X ball (assumes 1X additional weight in long end).

Equation: Assuming ball = 1 with variable application, and long end has additional 1 constant application, and counterweight located on shorter end, and counterweight is designed to direct ball on opposite end up slight supporting incline before ball applies leverage,

Unified Counterweight Mass Formula = Min Lvg + 1 > (Max Lvg / 2) + 1.

Modular Buoyancy Device 1

MODULAR BUOY DEVICE 1 (ModBD1)

LVG   COUNTERBUOY MAX OU

STATS DO NOT YET ACCOUNT
OF WATER RESISTANCE

1.75-2.25:1 >1.125 <1.75 <133.75% AVG

DIFF 1 COUNTED AS MASS
(DIFF -1 BUOYANCY)

PRIMARY BUOY PICTURED BELOW IS 1X
PROPORTION OF BUOYANCY

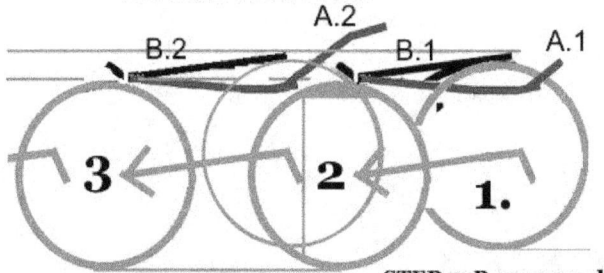

NATHAN LARKIN COPPEDGE

STEP 1: Counter-Buoy (not pictured, off to right) applies upward pressure, pushing 1X Buoy in position 1 using upside-down lever A.1. Buoy moves to position 2 up track B.1

STEP 2: Buoy 1 reaches position 2 where it applies buoyancy using leverage of 1.75X. The buoy then reaches a position of upper support (it is a buoy). It is then acted on by another counter-buoy by upside-down lever A.2. The primary buoy then moves to position 3 using support from upside-down track B.2 continuing.

[Modular Buoyancy Device]

Date of Invention: July 19, 2021

Precedents: 1stFPPMM, August 2016

Rating: 133.75% Conventional OU

Leverage: 1.75 to 2.25 : 1

Counterbuoy Buoyancy: > 1.125 < 1.75

Difference: -1 (mass on long end counts towards counterbuoy buoyancy)

Moving Rocks

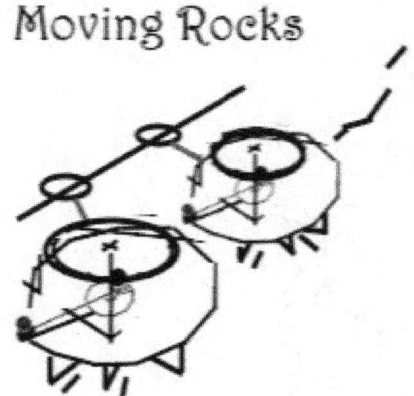

- Date of Invention: by June 19, 2019.
- State of affairs: Save for later.
- Rating: < 137.5% conventional Over-Unity
- Leverage = 1:0.5
- Counterweight Mass = >1X to <1.75X
- Maximum Gradient: approx < 18 degrees (not calculated).

Perpetual Motion Water Park Ride

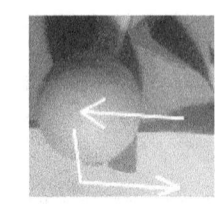

NEW
Escher Machine Principle:
For upwards motion:
Backboard curves from
about 25 degrees to
about 15 degrees.
Wedge is sloped on
the horizontal not
vertical. Experiment.

Nathan's Amazing Loop-the-Loop:

Visitor enters bubble-ball with
foam padding, is locked inside.

1. Bubble ball rolls through the loop-
the-loop using gravity and lower support.

Visitor splashes down into beginning
of Escher Machine to cushion fall.

The longer the Escher Machine
the better...

to return the ball
upwards
(see detail from
my working experiment).

Enclosed ball may now be opened,
with visitor safe due to foam padding and
predictable movements of the ball.
(Foam padding may be selected by visitors
prior to the ride).

Visitor will want to take the ride again!

Sponge Device

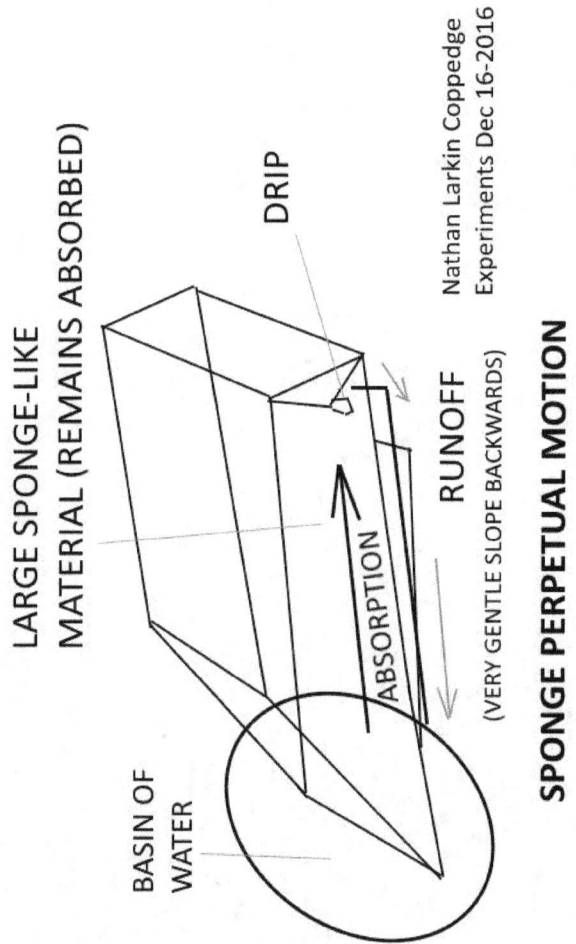

LARGE SPONGE-LIKE MATERIAL (REMAINS ABSORBED)

DRIP

Nathan Larkin Coppedge
Experiments Dec 16-2016

RUNOFF

ABSORPTION

(VERY GENTLE SLOPE BACKWARDS)

BASIN OF WATER

SPONGE PERPETUAL MOTION

It is thought this was thought of by Jen F

Supported Flying Machines

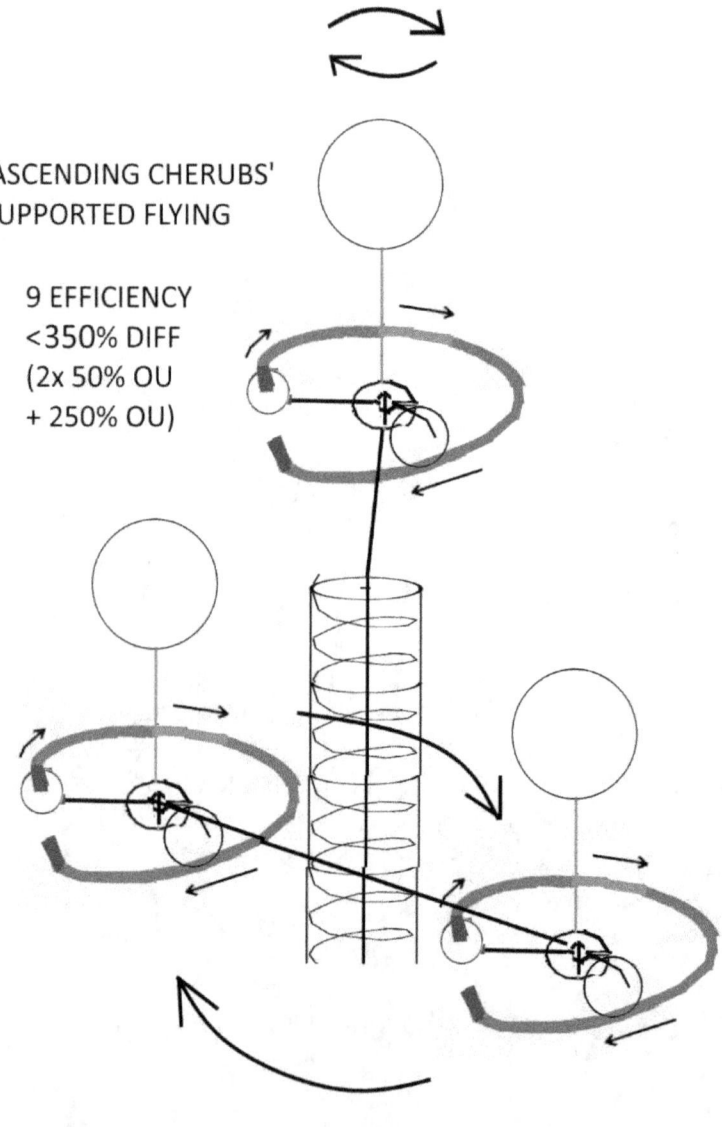

'ASCENDING CHERUBS'
SUPPORTED FLYING

9 EFFICIENCY
<350% DIFF
(2x 50% OU
+ 250% OU)

Swivel Buoy Device

SWIVEL-BUOYANCY DEVICE

Leverage: 2:1
Counterbuoy: >0 <1
Mobile Buoy: 1X
The structural mass
        and mobile
        buoy cancel
        out yielding
        very low
        downward
        resistance in
spite of leverage.

Nathan
Coppedge
2023-07-29

Tesla Valve

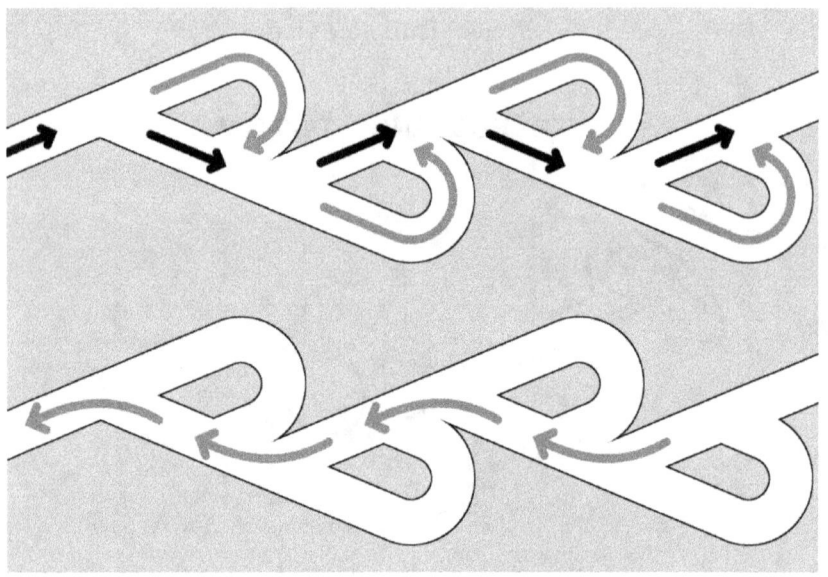

Above: Operation of a Tesla Valve detailed by CMG Lee

(Wikipedia)

Underwater Bubblevator

First known diagram: November 11, 2021

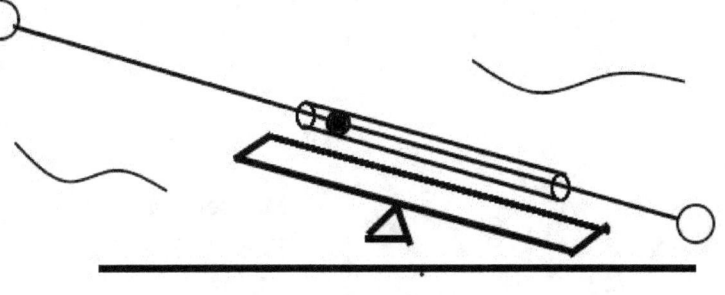

THE UNDERWATER BUBBLEVATOR

BY 2021

Rating: < 116.67 % Conventional OU

"Spaghetti" Leverage: 1 to 1.5 : 1

Center Mass Range: >2.75X <3X

Additional "spaghetti" buoyancy: 2

Estimated Maximum gradient: < 7.425 degrees.

Underwater Buoy Device 1

UNDERWATER BUOY DEVICE 1

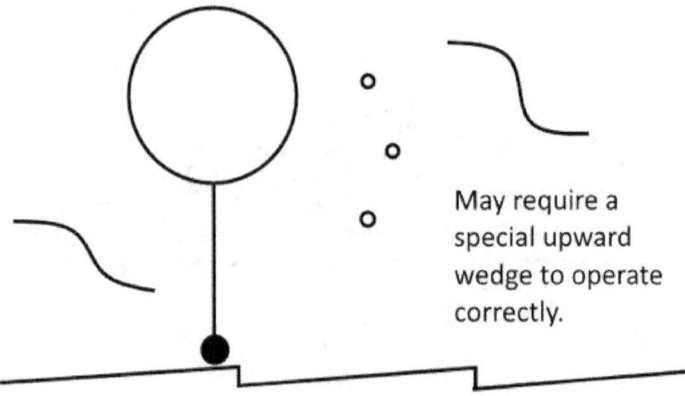

May require a special upward wedge to operate correctly.

> 50% to < 100% buoyancy with 1X mass

116.5% OU minus friction

...

Underwater Escher Machine

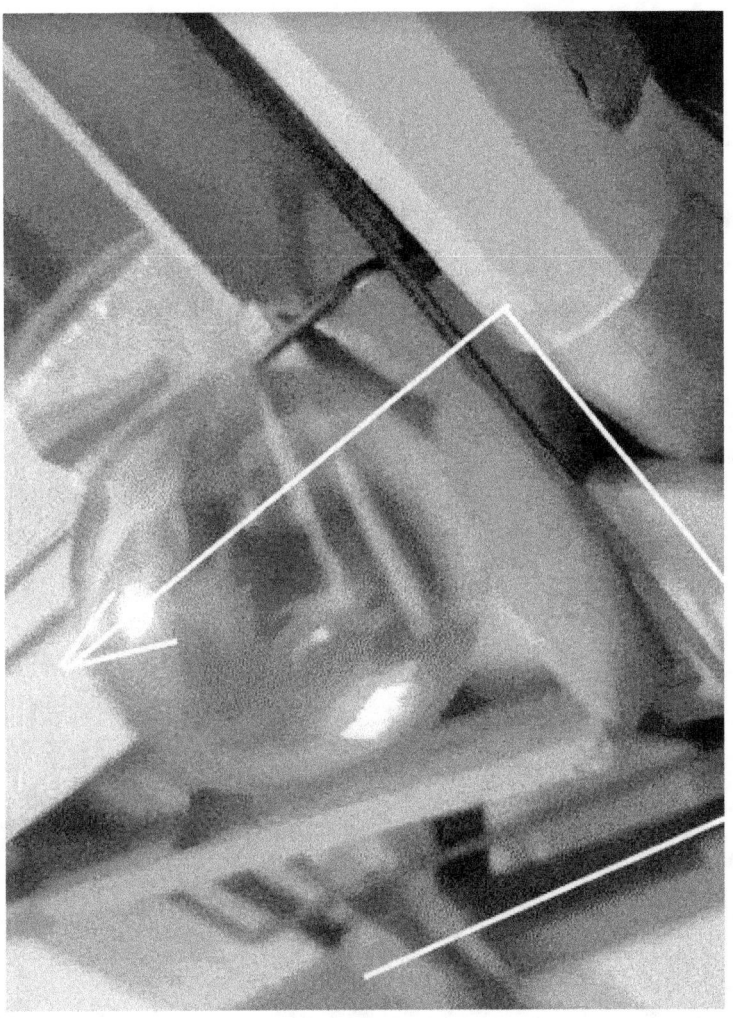

Upside Down Avant-Garde Escher Machine

This device uses a buoyant roller-buoy or similar with no counterweight and no counterbuoy upside down, using an upside-down Escher Machine to 'lift' the buoy downwards (using the Escher Principle).

Water Bailing Method 2

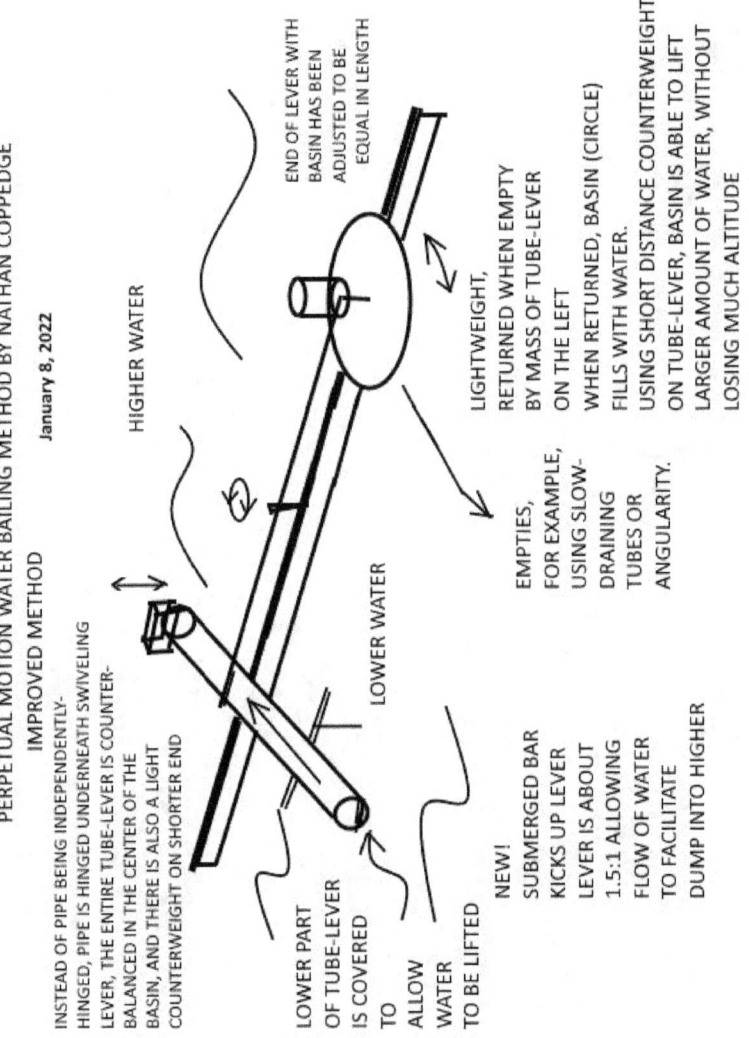

PERPETUAL MOTION WATER BAILING METHOD BY NATHAN COPPEDGE

January 8, 2022

IMPROVED METHOD

INSTEAD OF PIPE BEING INDEPENDENTLY-HINGED, PIPE IS HINGED UNDERNEATH SWIVELING LEVER, THE ENTIRE TUBE-LEVER IS COUNTER-BALANCED IN THE CENTER OF THE BASIN, AND THERE IS ALSO A LIGHT COUNTERWEIGHT ON SHORTER END

HIGHER WATER

END OF LEVER WITH BASIN HAS BEEN ADJUSTED TO BE EQUAL IN LENGTH

LOWER WATER

LIGHTWEIGHT, RETURNED WHEN EMPTY BY MASS OF TUBE-LEVER ON THE LEFT

WHEN RETURNED, BASIN (CIRCLE) FILLS WITH WATER.

USING SHORT DISTANCE COUNTERWEIGHT ON TUBE-LEVER, BASIN IS ABLE TO LIFT LARGER AMOUNT OF WATER, WITHOUT LOSING MUCH ALTITUDE

EMPTIES, FOR EXAMPLE, USING SLOW-DRAINING TUBES OR ANGULARITY.

LOWER PART OF TUBE-LEVER IS COVERED TO ALLOW WATER TO BE LIFTED

NEW! SUBMERGED BAR KICKS UP LEVER LEVER IS ABOUT 1.5:1 ALLOWING FLOW OF WATER TO FACILITATE DUMP INTO HIGHER

Water Lever

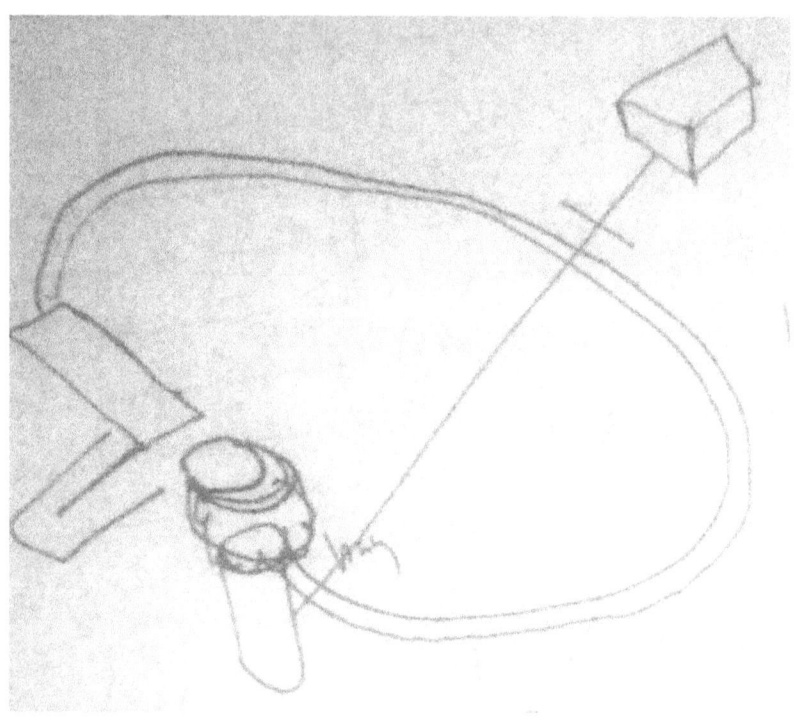

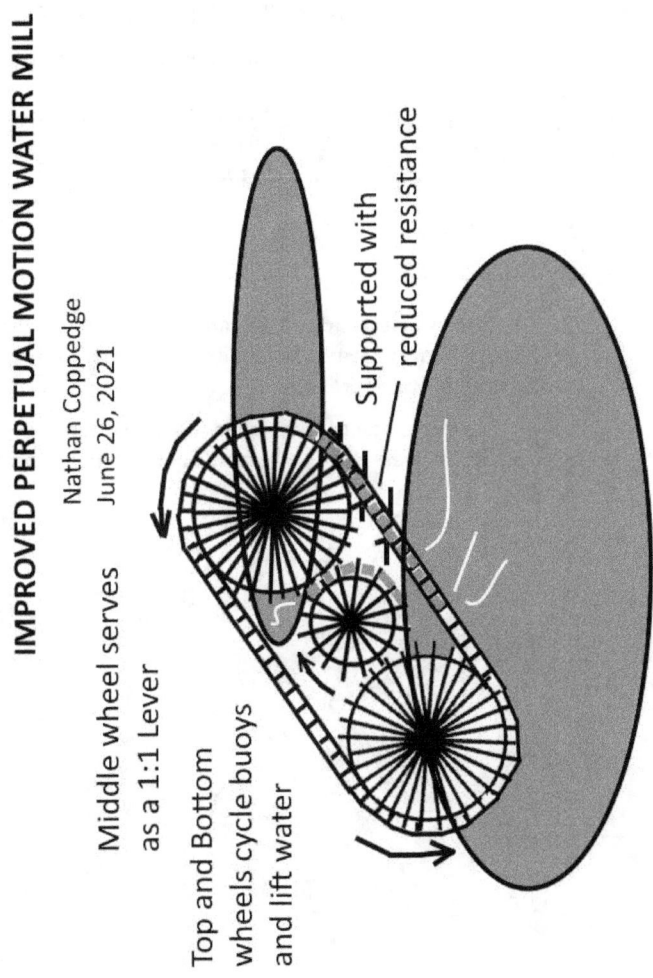

IMPROVED PERPETUAL MOTION WATER MILL

Nathan Coppedge   June 26, 2021

Middle wheel serves as a 1:1 Lever

Top and Bottom wheels cycle buoys and lift water

Supported with reduced resistance

Yan Yang's Buoyancy Torsion Wheel

These buoys fall, applying mass

BUOYANCY-TORSION WHEEL

Nathan Coppedge Jan 4, 2021

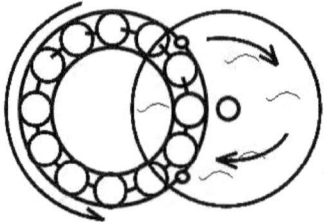

Thought to be a variation on designs attributed to the Chinese.

◯ = BUOY

Holes in top and base of tank allow water to cycle within the water-filled wheel without introducing great water pressure.

Entry-resistance is designed to be reduced due to the water being displaced back into the lower tank.

50 Great Flying and Underwater Perpetual Motion Machines

SUGGESTED READING

100 Great Perpetual Motion Machines

Bio

Nathan Coppedge or Nathan Larkin Coppedge (b.1982) is a philosopher, artist, inventor, poet, and member of the international honor society for philosophers. A prolific author with over 186 books published on Amazon, he is a perpetual motioneer, famous quotable, and internationally-selling Hyper-Cubist. A one-time member of Tesla Society UK online and PESWiki, and founder of many Facebook groups, he lives near Yale University.